Hurricane Katrina--WHY ?

Author: Olive Fuller

To order additional copies of this book, contact:
Xlibris Corporation
1-888-795-4274
www.Xlibris.com
Orders@Xlibris.com

At first we waited in anticipation. We boarded and battened down, filled up our tanks, checked our lanterns and candles and filled our cupboards with non-perishables. Computers were locked and schools were closed and we waited.

Then the gray clouds eerily marched in as the palm trees stood on tip-toes listening and waiting with great anticipation.

The shutters came down and so did the darkness, and we w-a-i-t-ed. Then" whoosh "whispered the palm trees as they began an anxious dance. W-W-H-O-O-O-S-H, you answered back as the raindrops started drumming on the roofs.

First you appeared quite harmless, lulling us into a false sense of safety. Then something made you very angry," WWW-HHH-OOO-SSSH "you shouted as you began pounding the roofs, battering the windows, throwing ficus trees, toppling light-posts, hurling decks and porches , and snapping shutters into lesser pieces.

You blew the lights out and plunged us into blackness. By candle light and lanterns, we huddled and listened as you furiously thrashed everything you found. We listened to your fury hoping you wouldn't bring terrible harm and injury.

Then He said, "Peace Be Still" and you obeyed and slowly left, leaving darkness and chaos and brokenness. Streets were blocked, stop-lights were down, houses became infernos. No one could go to their favorite places.

Days passed and the roads were cleared, as huge ficuses were asphalted: still the darkness lingered and the days and nights burned, but the palm trees waved in adoration and praise.

You left us and went back to the ocean, there you gathered speed and headed over to New Orleans. You broke the levees and flooded the Ninth Ward; scattering families and destroying memories.

We watched as grandmas hover on roof-tops and tree branches: waiting to be rescued even airlifted, buses, cars and trucks came to an ominous halt as busy highways, became holiday parking lots, people were moving but getting nowhere.

We watched and we waited as masses filled the superdome, leaving homes buried in watery mounds.

There they waited in pain and sadness, and hunger and death losing all they owned and dreading the unknown—They waited – for basic needs: water , food, a loving touch, and a place to call home again.

HURRICANE KATRINA, WHY DID
YOU BRING SO MUCH HEART-
ACHE AND PAIN ?